Project*Libre*
Practice Project

The Step by Step Sequence for Success

1x1MEDIA

By

Lisa A. Bucki

1x1 Media
Asheville, North Carolina
United States

Care has been taken to verify the accuracy of information in this book. However, the authors and publisher cannot accept responsibility for consequences from application of the information in this book, and makes no warranty, expressed or implied, with respect to its content.

Trademarks: Some of the product names and company names included in this book have been used for identification purposes only and may be trademarks or registered trade names of their respective manufacturers and sellers. The author and publisher disclaim any affiliation, association, or connection with, or sponsorship or endorsement by, such owners.

Do, Don't, and Tip icons designed by Freepik (contributor Gregor Cresnar).

ISBN 978-1-938162-13-8

© 2017 by 1x1 Media, LLC

email: info@1x1media.com

website : www.1x1media.com

Table of Contents

Project*Libre* Practice Project

The Step by Step Sequence for Success

"A good plan today is better than a perfect plan tomorrow"

~ Proverb

1

Introduction: Why ProjectLibre?

Project management, like any other professional discipline, requires using the proper tools to get the job done. You need project management software, and that's where ProjectLibre comes in. This open source (free) desktop project management software includes professional-grade project management features. With ProjectLibre, you can build a full project timeline and track the resources and costs needed to complete each task. ProjectLibre offers a variety of views, enables you to track your project's history through baselining, includes reports for sharing project progress, and more. If you want to move to a professional level of project management today, trust ProjectLibre desktop to get you there.

What This Book Provides—and Doesn't

ProjectLibre Practice Project provides a hands-on example for using the software to plan and track a basic project. Primarily geared for beginning users, the process followed in this book shows how to use ProjectLibre within the process and framework recommended for professional project management, Note that this is not a book about project management. To learn about project management, buy a book about project management or head over to the website for the Project Management Institute (www.pmi.org).

A Few Assumptions about the Software and You

The ProjectLibre team offers versions of its desktop software for Windows, Mac OS X, and Linux. Note that this version of ProjectLibre functions primarily as a single user, single project system. For the purposes of writing this book, I've assumed you have downloaded the appropriate version via the ProjectLibre website and have installed it on your system. I've also assumed that you are familiar with the basics for using your operating system and programs, and as a result, this book jumps right into coverage of ProjectLibre.

The screenshots for the figures in this book were taken at a screen resolution of 1024 x 768, with ProjectLibre's Windows version 1.7 on the Windows 10 operating system. I used the lower resolution so the figures would be easier to view on smaller ebook reader displays. (I have increased the size of some cropped screenshots to enhance readability.) If you have your system set to a higher screen resolution or are working on a different operating system, your screen will look somewhat different than the figures. However, the commands and options can all still be located as illustrated in this book.

I've also written the steps based on the Windows 10 operating system. In a limited number of instances, the instructions might differ slightly from what you need to do in your operating system. For example, when a step says to right-click, Mac OS X users would need to use the Control+click combination. I've assumed you are familiar enough with your operating system to substitute the appropriate action or shortcut when needed.

Book Organization, Elements, and Downloads

As noted earlier, ProjectLibre should be used according to a process defined for professional project management. For that reason, *ProjectLibre Practice Project* presents information in roughly the same order as the project management process:

1. Initiating

2. Planning

3. Executing

4. Monitoring and Controlling

5. Closing

As you follow the steps in this book to create the example project plan, the book will explain how the specific activities fit within the project management process.

When information requires special attention, I've flagged it for you as one of the following special elements:

Do: Highlights the correct or essential way to do something, or a step or method that you shouldn't leave out.

Don't: Identifies something that you should *not* do, either as a matter of following best practices or to avoid introducing problems into your project plan.

Tip: Provides hints, shortcuts, or additional details.

Files and Downloads

The final example project file shown in the figures is available for download. You can use it to test and practice various actions on your own.

Grab the file here:

> http://1x1media.com/ wp-content/ uploads/2017/03/
> System-Implementation-Final.pod_.zip

Note that this file is a compressed Zip file. Double-click the file to un-zip it.

Acknowledgments

My thanks go out to the founders of ProjectLibre, Marc O'Brien and Laurent Chretienneau. Their support and encouragement inspired me to move forward with this book as an independent project. Here's to continued success for them and the ProjectLibre team!

2

Road Map: Creating and Managing a Project

ProjectLibre Practice Project leads you through an example of how to use ProjectLibre to build and track a project plan. If you take the time to sit down and follow the steps presented here, you'll have a better sense of where ProjectLibre provides the greatest benefits to you as project manager and what you can expect in terms of using ProjectLibre to manage a live project plan in your organization. Here are the actions you'll perform within ProjectLibre, in order:

1. Make project plan file and custom calendar

2. Set the project start date and calendar (overall project parameters)

3. Add a list of tasks

4. Outline and link tasks to build the schedule

5. Add the list of resources needed to complete the project

6. Assign resources to the tasks in the schedule

7. Save the baseline snapshot of the plan

8. Track completed work

9. Communicate using views and reports

3

Understanding Project Initiation

Before you would dive in to the steps for building a project plan, you would work outside of ProjectLibre to complete what would serve as the Initiating phase of the project management process. This phase includes establishing the business case for the project, performing a feasibility study, developing the project charter, and identifying the project team and phases.

In the example project, which this book calls *System Implementation*, you will be playing the role of the leader of a team that needs to select and implement a new information system within your organization. You will set up the project plan file, add tasks, create the project schedule, and identify and assign resources. You will move on to save a record of the original plan, track work against the plan, and use views and reports to access information for control and communications.

Note that for the purposes of keeping the example project brief, it includes far fewer tasks than a real world system implementation project would typically require.

4

Creating the Project Plan File and Project Calendar

After you complete the Initiating phase for a project, you shift to the Planning phase by beginning to work in ProjectLibre. The first steps for creating a project plan include starting ProjectLibre, creating the file that will hold your project plan, and creating the custom calendar that the project will follow.

Working the calendar may seem like an odd place to start, but it's an essential step that can cause issues with the project schedule if done at a later time. Every organization has its own working hours and holidays (nonworking days). Many organizations follow a 40-hour work week with 8 a.m. to 5 p.m. working hours. Other companies work on a 24/7 schedule, or have shifts with various starting times. For ProjectLibre to schedule work correctly in your project plan, you have to tell the program what working schedule your organization or the team involved in the project follows. This is called setting the *base calendar* for the project plan. ProjectLibre comes with three built-in calendars, but none of these calendars has any holidays marked. So, *every* project plan file you create will require that you create a custom calendar reflecting scheduled nonworking days for your organization.

Because you can modify one of ProjectLibre's existing calendars, creating the custom calendar usually doesn't take long. We'll assume the fictional team members working on the example project all work a regular 40-hour work week. As a result, the steps instruct you to copy the Standard calendar in ProjectLibre and make changes to the calendar copy.

Begin by starting ProjectLibre and saving a new file:

1. Start ProjectLibre. The method for doing so will vary depending on your operating system and version. In Windows 10, click the **Start** button, scroll down the alphabetical list of programs to the P section, click **ProjectLibre**, and then **ProjectLibre** to start the ProjectLibre program.

   ☆ **Tip:** Of course, if you use ProjectLibre frequently in Windows 10, you can pin it to the Start menu or taskbar, or add a desktop shortcut for it.

   ☆ **Tip:** The first time you start ProjectLibre, the ProjectLibre License dialog box appears. Click the **I Accept** button to accept the license agreement. In the ProjectLibre Customer Information dialog box that appears, enter your email address in the **E-mail address** text box, and then click the **OK** button.

2. If the Tip of the Day dialog box appears, click the **Close** button.

3. In the Welcome to ProjectLibre dialog box, click the **Create Project** button. (If ProjectLibre was already open, you can click the **New** button in the **File** section of the **File** tab to create a new file.)

4. Type **System Implementation** in the **Project Name** text box of the New Project dialog box, and then click **OK**.

5. Click the **File** tab if it doesn't appear, and then click the **Save** button in the **File** section. (You also can click the **Save** button on the toolbar.) The Save dialog box appears.

> ☆ **Tip:** I'll also be summarizing how I refer to ribbon commands in some instances. If the text says, "Choose **Task>Views>Zoom In**," that means to click the **Task** tab on the ribbon, find the **Views** section, and click the **Zoom In** button in the section.

6. If the file name (which is the same as the project name specified in Step 4) doesn't appear automatically, type **System Implementation** in the **File name** text box, as shown in Figure 1. In this instance you are not changing the save location for the file. You are saving the file in the default *Documents* (Windows 10) folder on your system.

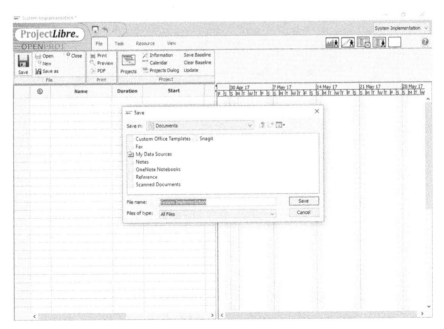

Figure 1 Name and save the project plan file.

7. Click **Save**. ProjectLibre displays the new file name in the title bar.

Now, make a copy of the Standard calendar in ProjectLibre and make changes to the custom calendar:

1. Choose **File>Project>Calendar**. The Change Working Calendar dialog box appears.

2. Click the **New** button in the lower-left corner of the dialog box.

3. In the New Base Calendar dialog box that appears, click the **Make a copy of the calendar** option button, make sure that the calendar to copy is selected from the accompanying drop-down list. If it's not, click the drop-down list and then click the calendar you want to copy, in this case, the **Standard** calendar as shown in Figure 2.

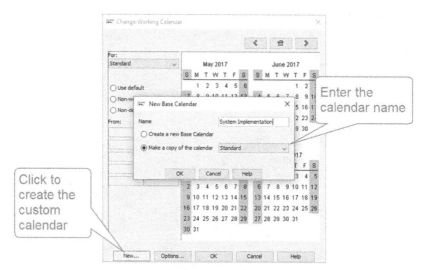

Figure 2 Copy the Standard calendar to create the custom calendar for the example project.

4. Type **System Implementation** into the **Name** text box. That is the name for your custom calendar.

5. Click **OK**. This takes you back to the Change Working Calendar dialog box. The For drop-down list now displays System Implementation. The changes you make next will be saved into the custom System Implementation calendar.

6. Use the right arrow button above the calendar in the dialog box to scroll right (forward) to **December 2020**.

    👍 **Do:** This book assumes you are working prior to December 2020. If you are working at a later time, adjust all dates to refer to similar dates in 2022 or beyond.

7. Click the date December **31** on the calendar.

8. Click the **Non-working time** option button at the left. This marks December 31, 2020 as a nonworking day (see Figure 3) in the custom System Implementation calendar.

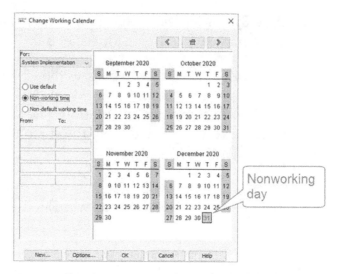

Figure 3 Mark a holiday using the Non-working time option button.

9. Scroll the calendar to **January 2021**.

10. Click January **1** on the calendar.

11. Click the **Non-working time** option button at the left. This marks January 1, 2021 (Figure 4) as a nonworking day.

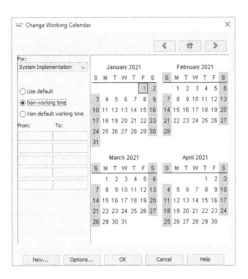

Figure 4 Setting January 1, 2021 as a nonworking day.

12. Click **OK**. ProjectLibre finishes saving the custom System Implementation calendar in the System Implementation file. The calendar isn't active in the file yet. You'll learn to apply the calendar in the next section.

👍 **Do:** You also can click the **Non-default working time** option and then make entries in the **From** and **To** text boxes to change the working hours for the selected day. (To select a day of the week for the entire calendar, click the day of the week at the top of one of the monthly calendar; your changes will then apply to all instances of that day of the week.) If you choose a calendar or working hours that are different from a standard 40-hour work week with 8 a.m.–5 p.m. (17:00 hours on the 24-hour clock) daily working hours five days a week, you also need to click Options in the Change Working Calendar dialog box. This opens the Duration Settings dialog box, where you should edit the **Hours per day**, **Hours per week**, and **Days per month** settings to match the calendar's working hours to ensure ProjectLibre schedules tasks correctly.

21

👍 **Do:** Be careful when creating your custom calendar. Each calendar you create is saved with the ProjectLibre program, and as of this writing, there is no way to delete a calendar.

5

Choosing Overall Project Parameters—Start Date and Calendar

When engaging in project managements, it makes sense that you would plan each project well in advance of the beginning of the work, allowing time for a conducting a thorough review of the plan and for receiving schedule and budget approvals.

When developing the plan, begin at the beginning. Specify in the plan file when the work will begin on the project—the project *start date*. After you enter the start date, ProjectLibre can build the schedule from that date automatically, saving you the trouble of entering other dates manually. Use the Project Information dialog box to specify the start date for each project plan file. You also use this dialog box to specify the calendar that the project will follow.

The start date may be based on a mandate from your boss, a team decision, the availability of a key vendor, or any other such factor. The example System Implementation project that you're creating will begin on December 1, 2020. The project also needs to follow the System Implementation custom calendar that you created earlier so that ProjectLibre takes into account the holidays that you marked. Choose both of those settings for your project now:

1. Still working in your System Implementation file in ProjectLibre, click **Information** in the **Project** section of the **File** tab. The Project Information dialog box appears.

2. Click the drop-down list arrow for the **Start** text box.

   In the pop-up calendar that appears, click the arrow buttons beside the month and year as many times as needed to display **December, 2020** at the top of the calendar (see Figure 5).

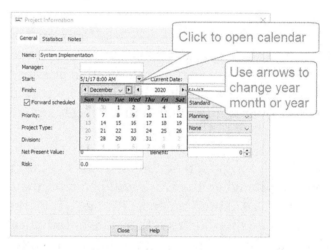

Figure 5 After displaying the year and month, click the desired start date on the calendar.

3. Click the **1** (for December 1, 2020) on the calendar. The calendar closes and **12/1/20 8:00 AM** appears in the Start text box in the dialog box.

   Don't: Although you can type dates directly into text boxes, using the pop-up calendar ensures that you won't specify a weekend date accidentally.

👍 **Do:** If the wrong time appears after you click December 1 on the calendar, do edit it to make sure that the Start text box displays **12/1/20 8:00 AM**. This will ensure that you get the same results as shown throughout this book as you follow along.

4. Click the **Base Calendar** drop-down list arrow and then click **System Implementation** (see Figure 6). Press **Tab** to finish the entry. This selects the custom calendar to be the calendar used to schedule the project within the project plan file.

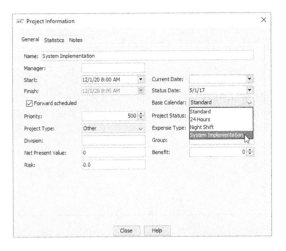

Figure 6 Use the Base Calendar drop-down list to assign the calendar the project will follow; in this case click the custom calendar you created earlier.

5. Click **Close**. ProjectLibre applies the specified start date and your custom calendar to the project plan file.

6. Save your work now by choosing **File>File>Save**.

Do: Before you download and open the finished example file, make sure you first follow the earlier steps for creating the custom calendar. Then use the Project Information dialog box to check to make sure it is correctly assigned to the finished example file. This can be necessary because, as noted earlier, custom calendars are saved within ProjectLibre.

You now have established the fundamental information that ProjectLibre needs to schedule the project plan correctly. You've told ProjectLibre that the System Implementation project should be scheduled to begin on December 1, 2020, and that the project will follow the System Implementation custom calendar that you created.

The next section leads you through the steps for listing the tasks to be completed during the course of the project.

6

Listing Project Tasks

After you establish the project start date and calendar, you can begin to identify the specific work that will be completed in order to produce the goals and deliverables for the project. As project manager, you identify how to break the goals and deliverables down into the individual steps or tasks required.

Each task should be as discrete as possible so that you can accurately track work against that completed task. For example, if a portion of the project requires running and monitoring five different diagnostic tests, then each test should be broken out as a separate task for better tracking accuracy.

👍 **Do:** Another approach to breaking tasks down to an appropriate level requires also thinking in advance about how many people you will assign to the task. Some tasks do require a team, but often you can break down a larger activity so that one or two people can handle each smaller part.

You will enter the project's list of tasks in the spreadsheet or sheet in the left pane of the Gantt view. Even though the default table in the sheet portion of the Gantt view contains multiple fields (columns), you will make entries for each task only in the Name and Duration fields. The Name field holds the name of each task, and the Duration field indicates how much time you think each task will take to complete from start to finish.

👍 **Do:** When entering tasks, make entries ONLY in the Name and Duration fields. Typing in dates for tasks limits some of ProjectLibre's capability to calculate the project schedule.

The example System Implementation project requires that you enter a number of tasks into the sheet at the left side of the Gantt view. Table 1 lists the entries you should make for each task in the Name and Duration columns, starting from the first row of the sheet. Type the Duration entries exactly as shown; if Table 1 does not have a Duration field entry for a task, leave the field blank for that task. ProjectLibre will enter an *estimated duration* (with a question mark) for those tasks. Note that the Duration field entries include the following duration labels or abbreviations:

- h for hours
- d for days
- w for weeks

Press Enter or Tab after you make each cell entry, and use the arrow keys to move around between cells as need. Note that when you enter a duration in hours (h) or weeks (w), ProjectLibre may change it to display in days. Make the entries in Table 1 in your copy of the System Implementation file now.

Table 1 Task Entries for the System Implementation File	
Name Field	**Duration Field**
Groundwork	
Identify affected departments	1w
Identify department representatives	1d
Planning meeting to discuss requirements	4h
Develop initial budget	1w
Selection	
Review existing workflow and software	2w
Identify systems (software and hardware) and vendors	2w
Review proposals	6h
Select solution and vendor	0
Planning	
White board workflow changes	1d
Role play new workflow	3d
Refine new workflow	2d
Develop implementation schedule	1w
Provide implementation plan to department representatives	2h
Resolve open issues or questions	3d
Documentation	
Develop workflow and software instructions	2w
Establish training schedule	2d
Perform employee training	3w
Launch	
Back up systems and data	2d
Setup and install	1w
Go live	4h
Optimize and troubleshoot	1w
Performance and budget audit	1w
Completion	0

When you finish, your file should look like Figure 7. Verify that it does, make any corrections needed, and choose **File>File>Save**. If needed, you can drag the divider at the right side of the Name column header to resize the column. Move on to the next section.

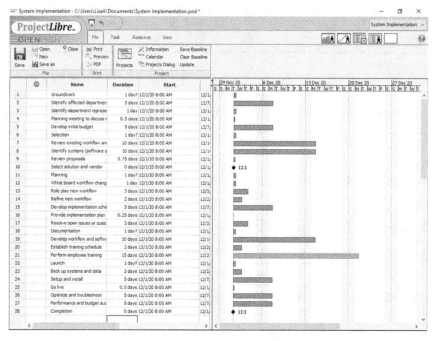

Figure 7 Type tasks in the System Implementation file.

⭐ **Tip:** The Gantt bar turns red for any task that is part of the *critical path* for the project. These are the tasks that have to finish on time for the project to finish on time.

👍 **Do:** If your project plan includes many multi-week tasks, in some cases you may want to use separate project plan files to track the detail work for those longer tasks. This will keep the overall project plan streamlined, while allowing for the management of the detail tasks broken out into separate smaller project files.

7

Organizing the Outline and Scheduling Tasks

Look closely at your project plan (or Figure 7). It may seem odd to see that:

- All the tasks start on the same date.

- That date is the project start date you specified earlier—December 1, 2020.

By default, ProjectLibre schedules every new task you add to begin on the project start date specified in the Project Information dialog box.

At this point, you should outline or organize tasks in the plan into logical groups of related tasks and/or chronological phases of the project. This process creates a type of task called a *summary task* that sums up or adds up the information about all the tasks in its group. When you entered the tasks from Table 1, you left the Duration field blank for some tasks. Those tasks will become summary tasks when you complete the steps in this section. Their estimated Duration field entries will be replaced with the summed Duration entries for the tasks (called *subtasks* or *detail tasks*) in each group.

Do: A question mark in a *Duration* field entry—as in 1 day?—indicates an estimated duration as noted earlier. If you are not confident about a Duration field entry for a task, type a question mark at the end of your entry to manually mark the duration as estimated. The question mark reminds you that you may need to revisit and finalize that Duration field entry at a later time. In between, you can consult with others such as other project managers or subject matter experts in your organization or vendors to arrive at a more accurate estimate for the duration.

After you organize the list of tasks, you need to add relationships between tasks to tell ProjectLibre how to build out the actual task schedule dates.

Do: Organizing tasks before scheduling them works best as a practical matter. Scheduling requires creating relationships between tasks, so those tasks need to be in place before you can create the relationships.

Organizing the Project Using Outlining

To organize your list of tasks, you will use tools similar outlining in other programs. You will *indent* (demote) tasks to make them subtasks using the Indent button in the Task group of the Task tab. Each row number along the left side of the sheet is also the *task ID number* for the task in that row. The rest of this book refers to tasks by the row/task ID number.

Follow these steps to organize (or outline) the list of tasks:

1. Drag over the row numbers for **tasks 2 through 5** to select those tasks. A selection highlight appears over the rows.

2. Click the **Task** tab, and then click the **Indent** button in the **Task** group. The button has a right arrow icon on it. As shown in Figure 8, ProjectLibre immediately indents the tasks and identifies task 1, Groundwork, as a summary task. You can tell that it's a summary task because its name now appears in bold and has a minus icon (for collapsing the group). The summary task's Gantt bar has changed to a black, summary task Gantt bar. Finally, the task's duration has changed from 1 day?, an estimated duration, to **5 days**, the current total duration for the indented subtasks: from the Start date of the earliest subtask to the Finish date of the latest subtask.

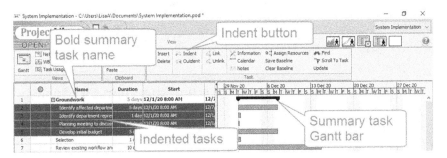

Figure 8 Indent tasks to outline the task list.

3. Select and indent **tasks 7 through 10**.

4. Select and indent **tasks 12 through 17**.

5. Select and indent **tasks 19 through 21**.

6. Select and indent **tasks 23 through 28**.

7. Click any cell to deselect tasks 23 through 28. Figure 9 shows how your project plan should looks with the outlining applied. You've identified five tasks as top-level summary tasks.

8. Choose **File>File>Save** to save the file.

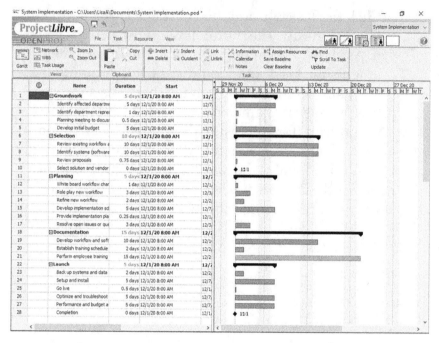

Figure 9 Indent subtasks to define the summary tasks.

👍 **Do:** Task 10, for which you entered a 0 duration, is a *milestone*. The Gantt view shows a milestone as a black diamond. The final task, 28, is also a milestone.

☆ **Tip:** After you've created a task that is not a milestone and have assigned a resource to it, you can convert the task to a milestone. This preserves the duration, schedule, and work information associated with the task, but displays it as a milestone on the Gantt chart. To make this change, double-click the **task's name** in the spreadsheet side of the view to display the Task Information dialog box, click the **Advanced** tab, click the **Display task as milestone** check box to check it, and then click **Close**.

In addition to outlining the list of tasks by indenting, you can use a variety of techniques to make further changes:

- You can select tasks within a summary group and indent them again. That is, the "outline" for your list of tasks can have multiple levels. Just make sure that the lower-level summary tasks are truly summaries, not tasks representing actual work. You always want the lowest-level tasks to represent actual work to be performed.

- You can select one or more tasks and then choose **Task>Task>Outdent** to move tasks up a level. After you do so, check carefully to make sure the nearby tasks have the correct indention level, as outdenting can affect the outline levels for tasks other than the ones you selected.

- You can insert a task within the list of tasks. Click in the row where you want to insert the new task, and then choose **Task>Task>Insert**. The new task row appears, and the task row where you clicked moves down.

- You also can select or click in a task row and use **Task>Task>Delete** to remove the task.

- You can tasks around in the list by cutting and pasting, just as in other applications. The **Clipboard** group of the **Task** tab has the typical click **Cut** and **Paste** buttons for removing and then inserting a task. Just be sure to select the entire task by clicking its row number. There's also a **Copy** button. If you copy a task and then paste it elsewhere in the list, be sure to change the task name, as it's not a best practice to have multiple tasks with the same task name in a project.

⭐ **Tip:** If you make a mistake when altering the task list, remember you can click the **Undo** button on the tool-bar or press **Ctrl+Z**. This can be especially handy if you accidentally delete a task, because ProjectLibre deletes the task immediately without asking you to confirm that action. The toolbar also has a **Redo** button to reinstate the action that you've undone.

Adding Links to Calculate the Project Schedule

After organizing the list of tasks, the next step is to indicate the relationships between tasks. You add a *link* between two tasks to establish the relationship between them. Linking enables ProjectLibre to calculate task schedules based on the sequence of the tasks. Adding all the necessary links in the tasks enables ProjectLibre to calculate the full project schedule. (This is why you only want to enter Name and Duration information for tasks. Linking will cause task dates to be calculated for you.)

Tasks in a project occur in a particular order. One task must finish so that the next task can begin, for example. By default, ProjectLibre creates *Finish-to-Start (FS) links,* which schedule tasks one after the other. The task with the driving schedule is the *predecessor,* and the task that the predecessor task's schedule drives or affects is called the *successor.*

Add links into the System Implementation file now to have ProjectLibre calculate the project schedule:

1. Drag over the task row numbers for **tasks 2 through 5** to select those tasks.

2. Click the **Link** button in the **Task** group of the **Task** tab. The button looks like a chain link. As shown in Figure 10, ProjectLibre links the tasks and changes

the Start and Finish dates for tasks 3 through 5 to reflect the new sequence in the schedule. (If needed, resize the Start and Finish field columns and drag the divider bar between the panes so that you can see the contents of the Start and Finish fields.)

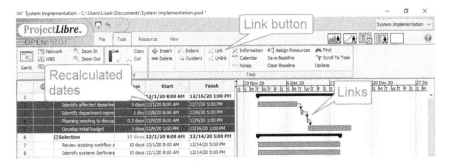

Figure 10 When you add links, ProjectLibre calculates the Start and Finish dates for linked tasks.

3. Click the task row number for **task** 5, press and hold the **Ctrl** key, click the task name for **task** 7, and then release the **Ctrl** key. Now click the **Link** button. ProjectLibre reschedules task 7 accordingly, but it also adjusts the Selection summary task to reflect the change to task 7's schedule.

👎 **Don't:** Summary tasks don't reflect specific work, so you should not link them. Only link the detail or sub-tasks so that ProjectLibre can accurately recalculate a summary task schedule when its individual tasks schedules change.

4. Select and link **tasks 7 through 10**.

5. Use the Ctrl+click method you learned in Step 3 to select **tasks 10 and 12** and then link them.

6. Select and link **tasks 12 through 17**.

7. Use Ctrl+click to select **tasks 17 and 19** and then link them.

8. Select and link **tasks 19 through 21**.

9. Use Ctrl+click to select **tasks 21 and 23** and then link them.

10. Select and link **tasks 23 through 26**.

11. Use Ctrl+click to select **tasks 25 and 27** and then link them. Linked subtasks tasks don't necessarily follow one after the other in the task sheet. You may need to link back to a task that's further up in the list if that task's schedule drives the schedule of one further down.

12. Finally, select and link **tasks 27 and 28**.

13. Look at the Finish field entry for task 28, the last task. ProjectLibre has used the linking information to calculate a finish date of 4/1/21 5:00 PM for that task and the project as a whole.

14. Click **task 23**, and then click the **Scroll To Task** button in the **Task** group of the **Task** tab. As shown in Figure 11, that button scrolls the right chart pane to display the Gantt bar for the selected task.

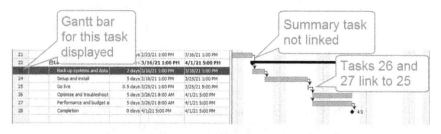

Figure 11 Review the dates after adding links.

15. Choose **File>File>Sa**ve to save the file.

☆ **Tip**: As shown in Figure 10, the **Ta**sk group of the **Ta**sk tab also includes **an Unli**nk button for removing the link between selected tasks.

Tweaking the Project Schedule

To keep the example simple, you have only added default Finish-to-Start (FS) links. ProjectLibre offers these additional link types:

- **Finish-to-Finish (FF)**: Means the linked tasks should finish on or about the same time.

- **Start-to-Finish (SF)**: Means the predecessor task's start time has a relationship to the successor task's finish point. (This is a rare circumstance.)

- **Start-to-Start (SS)**: Means the tasks start on or about the same time.

There also may be *lag time* (delay) associated with a link. For example, if you add 1 day of lag time with a Start-to-Start link, the successor task starts one day after the predecessor. Making a negative lag entry specifies *lead time*, or overlap in task schedules. You work with the link type and lag time entries in the Task Dependency dialog box that opens when you double-click a link line. Working with these settings can help you tweak the schedule to make it more realistic.

You also can use the Project Information dialog box to reschedule the entire project. For example, say you share a preliminary version of the plan with stakeholders, and the consensus is that the project needs to start earlier to hit a more aggressive completion date. Assuming you have used linking rather than typing in dates, you can reschedule the project simply by changing one date.

Follow these steps to practice working with more task scheduling in the example project:

1. Select task 20 and choose **Task>Task>Scroll To Task** to display its Gantt bars.

2. Double-click the **link line** to the right of the Gantt bar for **task 20**. The Task Dependency dialog box appears.

3. Change the entry in the **Lag** text box to **3 days** (or 3d), and then click **OK**. The link line extends and the Gantt bar for task 21 moves to the right to reflect the added lag time. (Figure 12 illustrates the changes to the bar and link line, as well as the change in the Task Dependency dialog box.)

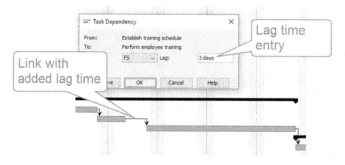

Figure 12 Add lag time to create a delay between tasks.

4. Scroll down if needed, and use the Ctrl+click method to select **tasks 25 and 27**, which you linked earlier. Choose **Task>Task>Unlink** to remove the link.

5. Select and link **tasks 26 and 27**.

6. With tasks 26 and 27 still selected, choose **Task>Task>Scroll To Task** to display the link between the tasks in the right pane.

7. Double-click the **link line** to the right of **task 26**. In the Task Dependency dialog box, click the **Type** drop-down list to open it, and then click **FF** for Finish-to-Finish. Change the Lag text box entry to **1 day** (or 1d), and then click **OK**. In this case the FF link indicates that task 26, Optimize and troubleshoot, must finish a day before task 27, Performance and budget audit. This makes sense, because optimization would need to be completed before the full audit process could be completed; the lag time specifies how much longer the audit process can last after the conclusion of optimization.

8. ProjectLibre has rescheduled the last several tasks based on the new lag time and links. Review the Finish field entry for **task 28**. It now shows a recalculated finish date of 4/7/21 5:00 PM, days later than the previously calculated 4/1/21 5:00 PM project finish date.

9. Scroll up to display **task 1**, if needed, and click its **Name** field. Choose **Task>Task>Scroll To Task** to return the Gantt bars for the early tasks back into view in the right pane.

10. Click the **File** tab to redisplay it, and then click **Information** in the **Project** group to reopen the Project Information dialog box. Click the **Statistics** tab. ProjectLibre displays calculated summary information about the whole project, as shown in Figure 13.

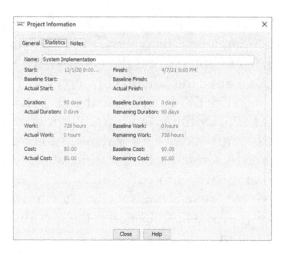

Figure 13 The Project Information dialog box displays a project summary and enables you to change the project start date.

11. Now say that based on the overall statistics, you want to move up the project finish date, which was pushed out based on the link changes you just made. Moving the project start date will pull up the finish date. Click the **General** tab if needed. Use the **Start** calendar to choose a date of **November 23, 2020**. Review that the Finish text box entry recalculates to 3/30/21 5:00 PM. Click **Close.**

12. Click the **Undo** button on the toolbar. (Shortcut: **Ctrl+Z**.) This returns the project to the previous 12/1/20 8:00 AM start date, with a finish date of 4/7/21 5:00 PM.

13. Choose **File>File>Save** to save the file before continuing to the next section.

Don't: The last two steps show why it is not a good practice to type start and finish dates for tasks. Allowing ProjectLibre to use the links you add to calculate dates means it can also recalculate the entire linked sequence of tasks based on earlier changes such as changing the project start date. Typing in dates removes that flexibility, so you would then have to retype dates to change the schedule. With linking, ProjectLibre only stops recalculating a task's start date after you mark work complete on a task.

8

Adding Project Resources

The System Implementation project will take three months to complete based on the initial task schedule that you've established. The next step in planning the project includes identifying who will complete the work and what materials, if any, will be used. The people and materials used to complete tasks are called the project *resources*. Entering resource information works like the process for entering tasks, but you work in a different view called Resources. Resources view offers several fields of information. You can make entries in all of or some of the fields, depending on your planning and tracking requirements. In the System Implementation project plan file, you will make entries in only the three most crucial fields to keep the example simple.

Follow these steps to add the list of resources in the System Implementation project plan file:

1. Click the **Resource** tab, and then click **Resources** in the **Views** group. (Note that you also can choose **View>Resource views> Resources**. However, using the Resource tab ensures that the other resource tools are visible.) The Resources view, which looks like a spreadsheet, appears.

2. Type the resource information listed in Table 2. You can press **Tab** and use the **arrow keys** to move between the cells, and simply skip any field not mentioned in the table. You also can click directly on a cell to make an entry in it.

45

You will need to scroll to the right using the horizontal scroll bar at the window bottom to see the Standard Rate and Base Calendar fields.

☆ **Tip:** If you use the Tab key to move into the Standard Rate field, you will need to use the **Backspace** key to delete the placeholder entry before typing the new entry. If you click directly on the cell, instead, you can then simply begin typing the desired entry.

☆ **Tip:** If you click the Base Calendar field cell for each resource, the System Implementation calendar should appear automatically. If it doesn't, open the drop-down list and click the System Implementation calendar.

Table 2 Resource Entries for System Implementation File		
Name	Standard Rate	Base Calendar
Maryann Melendez	50	System Implementation
Ray Cotter	60	System Implementation
Len Hu	60	System Implementation
Deepak Prasad	75	System Implementation
June Davis	75	System Implementation
Cam Jones	60	System Implementation

3. Now you need to add a *material resource* —a supply or item that will be used or consumed for one or more tasks. In the next blank cell in the Name field, type **USB Flash Drives** and press **Tab** twice. Click the selected cell in the Type field column to open the drop-down list, and then click **Material**. Press **Tab** twice to move to the Material Label field and type **5 Pack**. Press **Tab** four times, click to select the existing entry in the Standard Rate field, type **19.99**, and then press **Enter** or **Tab** to finish. (The Material Label and Standard Rate

entries correlate to describe the quantity and cost of the resource, in this case a 5 pack of USB Flash Drives costs $19.99.) Scroll back to the left so that the first field column is visible. Your finished Resources view entries should look like Figure 14.

4. Choose **File>File>Save** to save the file before continuing to the next section.

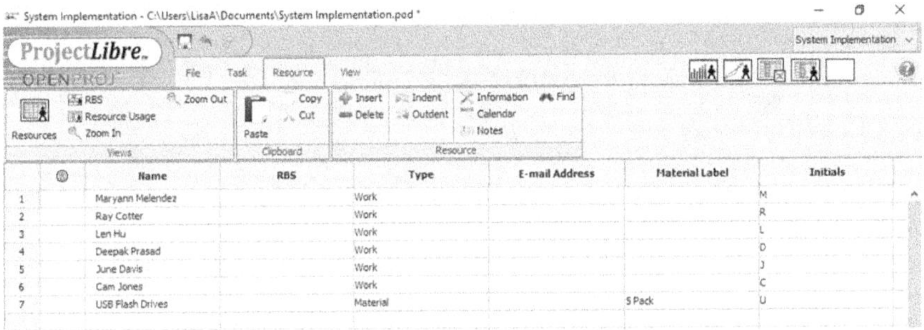

Figure 14 Use Resources view to add resource information.

👍 Do: When a resource follows a different calendar than the overall project calendar—such as when a person has vacation days—you should identify the resource's time off before proceeding with your planning. With the resource selected in the Resources view, click th**e Calen-dar** button in th**e Resour**ce group of the **Resource** tab. When the Change Working Calendar dialog box opens, the For drop-down list displays the name of the selected resource. Make the desired calendar changes for the resource, and then click **OK**.

9

Assigning Resources to Tasks

Next in the Planning phase for a project, you *assign* resources to tasks. You will work in Gantt view when making assignments, because you must use a view where tasks are visible.

When you assign the first resource(s) to a task, ProjectLibre does not change the Duration field entry you made for the task. Therefore, if you assign one resource to each task, the schedule will stay the same. However, ProjectLibre uses *effort driven scheduling*, a scheduling method that assumes that if you add more resources to a task, the task's duration should decrease because multiple resources working simultaneously results in the hours of work for the task being completed in a shorter timeframe. The decisions you make in assigning resources to tasks can dramatically affect the project schedule.

Follow the next steps to assign resources in the example System Implementation file:

1. Choose **Task>Views> Gantt**. (Once again, you could use the **View** tab to change views, but in this instance leaving the Task tab visible makes sense when working with tasks.)

2. Click **task 2**, Identify affected departments. Its Duration field entry is 5 days.

Don't: Assign resources only to detail tasks (the level at which the work actually occurs), not summary tasks. In rare exceptions you might assign a resource to a summary task to account for administrative work, but the norm means sticking with assigning resources only to detail tasks.

3. Choose **Task>Task>Assign Resources**. (If the Assign Resources button is inactive, clicking a cell in the spreadsheet portion of the view should reactivate it.) The Assign Resources dialog box appears.

4. Click **Maryann Melendez** in the **Name** list of the Assign Resources dialog box and then click **Assign**. Maryann Melendez's name appears beside the Gantt bar for task 2. Her resource name appears highlighted in the dialog box. Notice that the task retains its original duration of 5 days (see Figure 15).

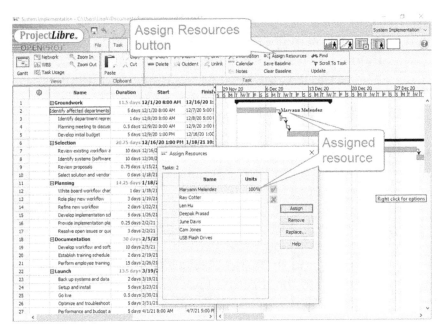

Figure 15 Make assignments with the Assign Resources dialog box.

5. Leaving task 2 selected, click **Len** Hu in the Assign Resources dialog box and then click **Assign**. Now both resource names appear beside the task bar in the Gantt chart. Due to effort-driven scheduling, the task's duration has decreased to 2.5 days.

6. Assume in this case that you do not want the task duration to decrease. Click **Len** Hu in the Assign Resources dialog box, if needed, and then click **Remove**. The task duration returns to 5 days. However, you want to assign both Maryann and Len to the task without affecting its duration. To do so, you need to remove Maryann from the task, too, so click **Maryann Melendez** in the Assign Resources dialog box, and then click **Remove**.

7. Now use Ctrl+click to select both **Maryann Melendez** and **Len Hu** in the Assign Resources dialog box and then click **Assign**. This time, both their names appear beside the task but the task retains its 5 days duration.

8. Click **task 3** in the task spreadsheet and then assign **Ray Cotter** to the task using the Assign Resources dialog box. Notice that you can leave the Assign Resources dialog box open and move between using it and the task spreadsheet.

9. Click **task 4** in the task spreadsheet, use Ctrl+click to select all resources except the USB Flash Drives resource, and then click **Assign**. This assigns all the resources to the meeting task without changing the task duration.

10. Click **task 5** in the task spreadsheet, use Ctrl+click to select **Deepak Prasad, June Davis**, and **Cam Jones**. Click **Assign**. This assigns all the resources to the meeting task without changing the task duration.

11. Click **task 7**, Review existing workflow and software. Say that you want to assign Ray Cotter to the task. You want to allow him the full two-week duration to complete the task, but you know he'll only be working on the task about 25% of the time. You need to indicate that ProjectLibre should not calculate the cost for his working hours as if he were working full time on the task. To do so, you will adjust the *units* for the assignment. Click the Name cell for **Ray Cotter** in the Assign Resources dialog box, double-click the **Units** cell to the right of the selected resource name, type **25**, and then press **Enter**. Figure 16 illustrates that Ray is now assigned to work on that task 25% of his time over two weeks.

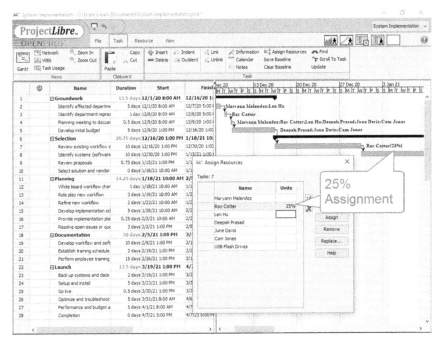

Figure 16 Change the Units setting for a to make the assignment part time.

Assign the resources as listed in Table 3 to the remaining tasks. Where the table lists multiple resources, use the Ctrl+click method to select all the resources and assign them at the same time. All of the assignments are full time. Tasks for which there are no assignments are summary tasks or milestones.

Table 3 Additional Assignments for the System Implementation File		
Task ID	Task Name	Resource(s)
8	Identify systems (software and hardware) and vendors	Deepak Prasad June Davis
9	Review proposals	Maryann Melendez Len Hu
12	White board workflow changes	All, except USB Flash Drives
13	Role play new workflow	June Davis Cam Jones
14	Refine new workflow	Deepak Prasad Cam Jones
15	Develop implementation schedule	Ray Cotter
16	Provide implementation plan to department representatives	Ray Cotter
17	Resolve open issues or questions	Maryann Melendez Ray Cotter
19	Develop workflow and software instructions	Ray Cotter Len Hu
20	Establish training schedule	Ray Cotter
21	Perform employee training	Ray Cotter Len Hu
23	Back up systems and data	Deepak Prasad June Davis
24	Setup and install	Deepak Prasad June Davis
25	Go live	Cam Jones
26	Optimize and troubleshoot	Deepak Prasad June Davis Cam Jones
27	Performance and budget audit	All, except USB Flash Drives

12. Assume that you've reviewed the plan and you want to accomplish task 15, Develop implementation schedule more quickly. Click **task 15** and assign **Cam Jones** to the task, too. Adding the second resource reduces the task's duration to 2.5 days.

13. You also realize that the training will require a USB flash drive for each employee trained, and 60 employees will be trained, meaning you will need 12 5 packs of USB flash drives. Click **task 21**, click Units cell to the right of the **USB Flash Drives** resource in the Assign Resources dialog box, type a **12**, and then press **Enter**. As shown in Figure 17, ProjectLibre assigns the specified number of drives to the task.

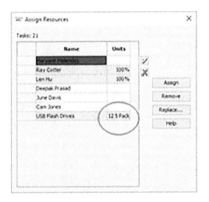

Figure 17 Enter the quantity for a material resource in the Units cell to assign the material resource.

14. Click the **Close (X)** button in its upper-right corner to close the Assign Resources dialog box.

15. Choose **File>File>Save**.

👍 **Do:** Because the example project file holds a relatively simple sequence of tasks, there is little chance that the assigned resources are overbooked. In the real world, you would want to examine the plan carefully to look for days when resources are assigned work hours exceeding 100% of availability. The Histogram sub-view, available in the Sub-views group of the View tab or via the Histogram button at upper-right, is useful for this purpose.

10

Saving the Baseline for the Plan

To finish the Planning phase of your project, you would perform a thorough review of the project plan to make sure that it is complete, accurate, and realistic. You would check assignments to ensure that the resources have the hours of availability to handle the tasks assigned, and you would circulate the plan and budget to secure the necessary sign-offs. After you and other stakeholders have given approval to the plan, the project moves from the Planning phase to the Executing phase, during which the team completes the tasks for the project.

To handle the Managing and Controlling activities for the project as Executing progresses, you as project manager need to see your progress versus the original plan. In ProjectLibre, you save the original plan as the **baseline**, or snapshot of the original schedule and budget. Later, as task schedules change based on actual work, ProjectLibre will be able to show you how the actual schedule varies from the original schedule.

Follow these steps to save the baseline:

1. Click **any cell** in task 1 to move back to the beginning of the System Implementation file, and then choose **Task>Task>Scroll To Task** to display its Gantt bars in the right portion of the view.

2. Still on the **Task** tab, click **Save Baseline** in the **Task** group.

3. In the Save Baseline dialog box (see Figure 18), make sure that the Baseline and Entire Project options are selected, and then click **OK** to save the baseline.

Figure 18 Save the baseline to track against the original plan.

4. Choose **File>File>Save** to save the System Implementation file. In this case, saving also preserves the baseline information in the file.

The Gantt chart appearance changes slightly after you save the baseline, with a second thin Gantt bar appearing for each of the summary and detail tasks. The top original bar represents the current schedule for the task. The bottom thin gray bar shows the baseline (original) schedule for the task.

11

Tracking Completed Work for Tasks

After completing the plan and saving the baseline, work begins on the project. In the real world, you would establish regular communications between yourself and all the resources to gather information about the amount of work completed on each task. You would then enter the information about work completed into the project plan so that ProjectLibre can calculate data such as the hours of work completed for the project as a whole and the portion of the budget now expended. You also would make adjustments to task schedules as needed so that ProjectLibre can recalculate the schedules for linked tasks accordingly.

The Update Tasks dialog box provides fields that enable you to enter actual completion information and see calculations such as actual versus remaining duration. To enter completion percentages, you also can use the Task Information dialog box; at the bottom of its General tab, you can compare the Baseline Start and Baseline Finish for the tab to the current Start and Finish scheduled for the task. Both dialog boxes include a Percent Complete text box where you specify the task completion percentage.

Follow these steps to track part of the work completed for the example System Implementation project:

1. Click **task 2**, Identify affected departments.

2. Choose **Task>Task>Update**. Select the entry in the **Percent Complete** text box, type **100**, and then click **Close**. (If you need to use the Delete or Backspace key to delete a current entry in the Percent Complete field, do so as needed throughout the steps.)

3. With task 2 still selected, choose **Task>Task>Update** again. As shown in Figure 19, a few changes occur in the Update Tasks dialog box. Actual Start and Actual Finish dates have been entered for the task. In Gantt view in the chart at the right, the top Gantt bar for the task has a black bar all the way through the middle, indicating that it is 100% complete. Plus, a green check mark appears in its Indicators field at far left. Click the **Close** button.

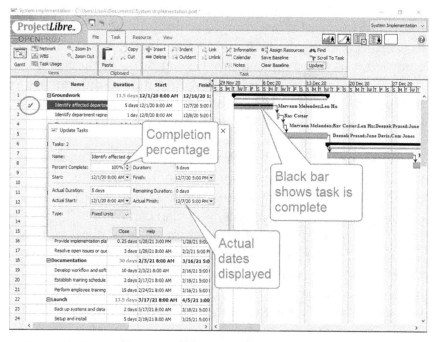

Figure 19 Mark work as complete on a task.

☆ **Tip:** You can update multiple tasks with the same completion percentage by selecting them before you open the Update Tasks dialog box.

4. Click **task 3**, Identify department representatives. This task is finished, but it started and finished a day later than scheduled, so you need to enter the actual start and finish dates that occurred. Choose **Task>Task>Update**. In the Actual area of the Update Task dialog box (refer to Figure 19), type **12/9/20** into both the Actual Start and Actual Finish text boxes, pressing **Tab** to finish each entry. (You also could use the text box drop-down calendars to specify the dates, if desired.) Click **Close**. As shown in Figure 20, the Gantt chart changes to reflect the delayed completion. The top Gantt bar for task 3 and the one for each subsequent linked task move to reflect the fact that task 3 occurred later than originally scheduled. The gray baseline bars remain in their original positions, comparing the baseline schedule to the actual and current task schedules.

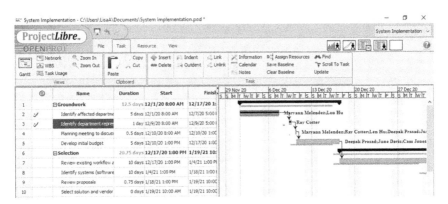

Figure 20 The Gantt bars now illustrate baseline versus actual and current dates.

☆ **Tip:** ProjectLibre tracks three sets of dates for each task: the saved *baseline* dates, the evolving *current* dates, and the *actual* dates based on when work was completed. Marking some work as complete on a task sets the Actual Start date, and marking the task as 100% complete sets the Actual Finish date. Once an actual date has been set, ProjectLibre will no longer reschedule the corresponding current dates.

5. Select **tasks 4 and 5**, open the Update Tasks dialog box, enter **100** in the **Percent Complete** text box, and then click **Close**.

6. Select **task 6**, Review existing workflow and software, open the Update Tasks dialog box, enter **25** in the **Percent Complete** dialog box, and then click **Close**. Only a portion of the task's Gantt bar is marked with a black bar to indicate the completion percentage.

7. Assume that that's all the work you need to track for now. Choose **File>File>Save** to save the System Implementation file.

12

Sharing Results through Views and Reports

Onscreen display and printouts of various views and reports can be useful tools during the Monitoring and Controlling phase of managing the project. They also are the essential documentation for the Closing phase of managing the project, which includes activities such as:

- Releasing deliverables to customers and stakeholders

- Documenting the project and sharing documentation with stakeholders

- Closing contracts and paying outstanding invoices or fees

- Performing reviews to measure success and identify lessons learned

ProjectLibre offers a variety of views which you can display via the View tab. To share the information onscreen in a view, you can print the view. ProjectLibre prints whatever view currently appears onscreen.

Print your file now, which is still in Gantt view, for example:

1. With the System Implementation file still onscreen, click **File** and then click **Preview** in the **Print** group. A preview view of the file appears onscreen.

2. Click the right arrow button at the top of the preview to scroll to page 2 of the printout. This page shows the Gantt bars for tasks, as shown in Figure 21.

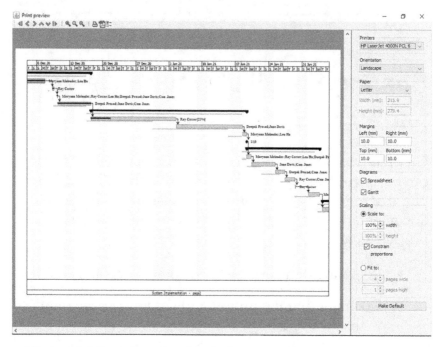

Figure 21 Preview the file before printing.

3. Review the print setup choices in the pane at the right, making any changes as needed.

4. Click the **Print** button on the toolbar, and then click **OK** in the Print dialog box to print the view and close the print preview. (Note that you could click **Print** in the **Print** group of the **File** tab to print without displaying the Print Preview.)

5. Click the preview window **Close (X)** button to close the window if needed.

Do: To create a paperless archive of your project documents, display each view as desired, and then choose **File>Print>PDF**. The Save dialog box opens, with the PDF (*.pdf) file format automatically selected. Edit the file name and save location as needed, and then click **Save**.

ProjectLibre also offers a Project Details report view that presents summary information about the currently opened project, as well as reports named Resource Information, Task Information, and Who Does What. Follow these steps to display and print a report of the information in the current project file:

1. Choose View>Other views> Report. The report appears onscreen

2. Open the Report drop-down list at the top of the preview, and click Who Does What. Then open the Columns drop-down list and click Tasks Assigned. The report appears in a print preview (see Figure 22).

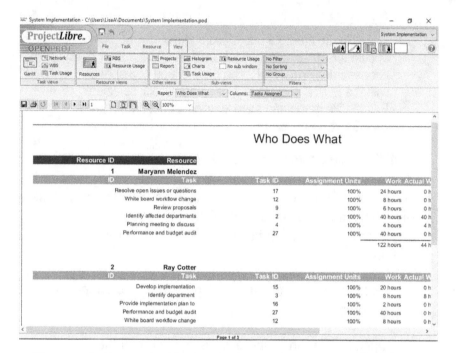

Figure 22 Reports display information in an attractive layout.

3. To print the report, click the **Print** button above the preview, change settings as needed, and then click **OK**.

4. Click **View** and then click **Gantt** in the **Task views** group.

5. Save the System Implementation file, and then click the **Close (X)** button at upper-right to close ProjectLibre.

Review

If you've followed the steps in this guide, you've now created and managed a basic project plan in ProjectLibre. You've now seen how to set up and choose a project calendar, specify a project start date, list tasks, outline and link tasks, list resources, assign resources, save the baseline, track work, and print a view or report. We hope the skills you've learned provide a strong foundational knowledge of ProjectLibre and increase your success using the program to manage your future projects.

Thank You

This concludes the *ProjectLibre Practice Project*. We hope you find our content and supporting tools useful for your journey.

We are always looking for feedback on our tools. If you have comments, feedback, or corrections, please send us a note.

info@1x1media.com

http://www.1x1media.com

###

www.ingramcontent.com/pod-product-compliance
Lightning Source LLC
LaVergne TN
LVHW052321060326
832902LV00023B/4524